MATH
WORKBOOK

EMMA.SCHOOL

ADDITION

1. $\begin{array}{r} 5 \\ + 25 \\ \hline \end{array}$
2. $\begin{array}{r} 14 \\ + 4 \\ \hline \end{array}$
3. $\begin{array}{r} 14 \\ + 21 \\ \hline \end{array}$
4. $\begin{array}{r} 18 \\ + 33 \\ \hline \end{array}$

5. $\begin{array}{r} 23 \\ + 25 \\ \hline \end{array}$
6. $\begin{array}{r} 35 \\ + 17 \\ \hline \end{array}$
7. $\begin{array}{r} 7 \\ + 6 \\ \hline \end{array}$
8. $\begin{array}{r} 8 \\ + 28 \\ \hline \end{array}$

9. $\begin{array}{r} 11 \\ + 34 \\ \hline \end{array}$
10. $\begin{array}{r} 14 \\ + 36 \\ \hline \end{array}$
11. $\begin{array}{r} 7 \\ + 20 \\ \hline \end{array}$
12. $\begin{array}{r} 31 \\ + 6 \\ \hline \end{array}$

13. $\begin{array}{r} 33 \\ + 37 \\ \hline \end{array}$
14. $\begin{array}{r} 10 \\ + 3 \\ \hline \end{array}$
15. $\begin{array}{r} 36 \\ + 26 \\ \hline \end{array}$
16. $\begin{array}{r} 19 \\ + 38 \\ \hline \end{array}$

17. $\begin{array}{r} 20 \\ + 40 \\ \hline \end{array}$
18. $\begin{array}{r} 28 \\ + 21 \\ \hline \end{array}$
19. $\begin{array}{r} 18 \\ + 7 \\ \hline \end{array}$
20. $\begin{array}{r} 12 \\ + 12 \\ \hline \end{array}$

1. 12
    + 16
    ....

2. 37
    + 37
    ....

3.   6
    + 37
    ....

4. 38
    + 7
    ....

5. 19
    + 27
    ....

6. 11
    + 5
    ....

7. 22
    + 3
    ....

8.   9
    + 23
    ....

9. 15
    + 33
    ....

10. 17
    + 33
    ....

11. 29
    + 26
    ....

12. 14
    + 36
    ....

13. 20
    + 10
    ....

14. 32
    + 28
    ....

15.   1
    + 4
    ....

16. 27
    + 10
    ....

17. 36
    + 35
    ....

18.   1
    + 13
    ....

19. 27
    + 9
    ....

20. 33
    + 31
    ....

1.　　6
　+ 11
....

2.　　7
　+ 16
....

3.　　40
　+ 2
....

4.　　9
　+ 28
....

5.　　3
　+ 37
....

6.　　35
　+ 7
....

7.　　1
　+ 26
....

8.　　30
　+ 27
....

9.　　21
　+ 12
....

10.　　7
　+ 18
....

11.　　22
　+ 29
....

12.　　6
　+ 24
....

13.　　33
　+ 26
....

14.　　18
　+ 19
....

15.　　7
　+ 15
....

16.　　10
　+ 26
....

17.　　31
　+ 36
....

18.　　26
　+ 33
....

19.　　36
　+ 30
....

20.　　3
　+ 29
....

1. $\begin{array}{r} 24 \\ + 29 \\ \hline \end{array}$
2. $\begin{array}{r} 21 \\ + 28 \\ \hline \end{array}$
3. $\begin{array}{r} 27 \\ + 10 \\ \hline \end{array}$
4. $\begin{array}{r} 16 \\ + 5 \\ \hline \end{array}$

5. $\begin{array}{r} 40 \\ + 37 \\ \hline \end{array}$
6. $\begin{array}{r} 31 \\ + 32 \\ \hline \end{array}$
7. $\begin{array}{r} 9 \\ + 18 \\ \hline \end{array}$
8. $\begin{array}{r} 24 \\ + 22 \\ \hline \end{array}$

9. $\begin{array}{r} 28 \\ + 19 \\ \hline \end{array}$
10. $\begin{array}{r} 2 \\ + 29 \\ \hline \end{array}$
11. $\begin{array}{r} 25 \\ + 12 \\ \hline \end{array}$
12. $\begin{array}{r} 31 \\ + 6 \\ \hline \end{array}$

13. $\begin{array}{r} 10 \\ + 19 \\ \hline \end{array}$
14. $\begin{array}{r} 31 \\ + 30 \\ \hline \end{array}$
15. $\begin{array}{r} 22 \\ + 10 \\ \hline \end{array}$
16. $\begin{array}{r} 28 \\ + 14 \\ \hline \end{array}$

17. $\begin{array}{r} 7 \\ + 28 \\ \hline \end{array}$
18. $\begin{array}{r} 6 \\ + 6 \\ \hline \end{array}$
19. $\begin{array}{r} 20 \\ + 14 \\ \hline \end{array}$
20. $\begin{array}{r} 40 \\ + 13 \\ \hline \end{array}$

1.   43
   + 14

2.   39
   + 19

3.   21
   + 39

4.   36
   + 28

5.   60
   + 37

6.   26
   + 30

7.   31
   + 2

8.   13
   + 33

9.   13
   + 15

10.   65
   + 8

11.   58
   + 2

12.   30
   + 37

13.   46
   + 8

14.   39
   + 12

15.   33
   + 30

16.   54
   + 19

17.   12
   + 14

18.   16
   + 16

19.   48
   + 32

20.   58
   + 20

1.   17
  + 5

2.   12
  + 36

3.   41
  + 5

4.   57
  + 32

5.   23
  + 20

6.   47
  + 6

7.   25
  + 14

8.   6
  + 19

9.   18
  + 10

10.   44
  + 3

11.   61
  + 39

12.   37
  + 5

13.   67
  + 26

14.   33
  + 36

15.   61
  + 23

16.   28
  + 24

17.   19
  + 33

18.   56
  + 31

19.   60
  + 18

20.   49
  + 24

1. $\begin{array}{r} 25 \\ + 18 \\ \hline \end{array}$
2. $\begin{array}{r} 44 \\ + 4 \\ \hline \end{array}$
3. $\begin{array}{r} 51 \\ + 30 \\ \hline \end{array}$
4. $\begin{array}{r} 6 \\ + 11 \\ \hline \end{array}$

5. $\begin{array}{r} 47 \\ + 19 \\ \hline \end{array}$
6. $\begin{array}{r} 26 \\ + 8 \\ \hline \end{array}$
7. $\begin{array}{r} 26 \\ + 12 \\ \hline \end{array}$
8. $\begin{array}{r} 55 \\ + 15 \\ \hline \end{array}$

9. $\begin{array}{r} 35 \\ + 29 \\ \hline \end{array}$
10. $\begin{array}{r} 29 \\ + 37 \\ \hline \end{array}$
11. $\begin{array}{r} 9 \\ + 2 \\ \hline \end{array}$
12. $\begin{array}{r} 36 \\ + 12 \\ \hline \end{array}$

13. $\begin{array}{r} 7 \\ + 3 \\ \hline \end{array}$
14. $\begin{array}{r} 17 \\ + 40 \\ \hline \end{array}$
15. $\begin{array}{r} 37 \\ + 18 \\ \hline \end{array}$
16. $\begin{array}{r} 28 \\ + 17 \\ \hline \end{array}$

17. $\begin{array}{r} 22 \\ + 21 \\ \hline \end{array}$
18. $\begin{array}{r} 19 \\ + 30 \\ \hline \end{array}$
19. $\begin{array}{r} 57 \\ + 9 \\ \hline \end{array}$
20. $\begin{array}{r} 7 \\ + 38 \\ \hline \end{array}$

1. $\begin{array}{r} 36 \\ + 4 \\ \hline \end{array}$
2. $\begin{array}{r} 51 \\ + 10 \\ \hline \end{array}$
3. $\begin{array}{r} 7 \\ + 39 \\ \hline \end{array}$
4. $\begin{array}{r} 62 \\ + 34 \\ \hline \end{array}$

5. $\begin{array}{r} 2 \\ + 2 \\ \hline \end{array}$
6. $\begin{array}{r} 38 \\ + 12 \\ \hline \end{array}$
7. $\begin{array}{r} 44 \\ + 18 \\ \hline \end{array}$
8. $\begin{array}{r} 67 \\ + 34 \\ \hline \end{array}$

9. $\begin{array}{r} 18 \\ + 24 \\ \hline \end{array}$
10. $\begin{array}{r} 4 \\ + 31 \\ \hline \end{array}$
11. $\begin{array}{r} 4 \\ + 7 \\ \hline \end{array}$
12. $\begin{array}{r} 36 \\ + 23 \\ \hline \end{array}$

13. $\begin{array}{r} 42 \\ + 29 \\ \hline \end{array}$
14. $\begin{array}{r} 24 \\ + 4 \\ \hline \end{array}$
15. $\begin{array}{r} 33 \\ + 37 \\ \hline \end{array}$
16. $\begin{array}{r} 58 \\ + 19 \\ \hline \end{array}$

17. $\begin{array}{r} 16 \\ + 16 \\ \hline \end{array}$
18. $\begin{array}{r} 32 \\ + 35 \\ \hline \end{array}$
19. $\begin{array}{r} 38 \\ + 22 \\ \hline \end{array}$
20. $\begin{array}{r} 58 \\ + 12 \\ \hline \end{array}$

1.   42
  + 44

2.   44
  + 34

3.   15
  + 36

4.   33
  + 27

5.   38
  + 11

6.   39
  + 36

7.   20
  + 35

8.   60
  + 68

9.   19
  + 88

10.   28
  + 60

11.   18
  + 23

12.   22
  + 79

13.   66
  + 24

14.   66
  + 85

15.   59
  + 4

16.   37
  + 78

17.   47
  + 45

18.   36
  + 14

19.   49
  + 7

20.   14
  + 26

1. 38
    + 41
    ⎯⎯⎯

2. 39
    + 39
    ⎯⎯⎯

3. 50
    + 55
    ⎯⎯⎯

4. 57
    + 60
    ⎯⎯⎯

5. 66
    + 62
    ⎯⎯⎯

6. 17
    + 8
    ⎯⎯⎯

7. 65
    + 3
    ⎯⎯⎯

8. 64
    + 68
    ⎯⎯⎯

9. 44
    + 19
    ⎯⎯⎯

10. 39
    + 47
    ⎯⎯⎯

11. 12
    + 51
    ⎯⎯⎯

12. 69
    + 76
    ⎯⎯⎯

13. 60
    + 41
    ⎯⎯⎯

14. 51
    + 17
    ⎯⎯⎯

15. 51
    + 15
    ⎯⎯⎯

16. 61
    + 85
    ⎯⎯⎯

17. 47
    + 22
    ⎯⎯⎯

18. 59
    + 29
    ⎯⎯⎯

19. 64
    + 24
    ⎯⎯⎯

20. 30
    + 71
    ⎯⎯⎯

1. 199 + 76	2. 98 + 43	3. 157 + 68	4. 39 + 79
5. 16 + 26	6. 161 + 78	7. 116 + 77	8. 60 + 20
9. 116 + 46	10. 161 + 76	11. 115 + 30	12. 13 + 45
13. 116 + 80	14. 61 + 12	15. 183 + 15	16. 150 + 19
17. 112 + 49	18. 123 + 86	19. 198 + 30	20. 88 + 61

1. $96 + 20$

2. $192 + 87$

3. $98 + 64$

4. $71 + 15$

5. $56 + 35$

6. $36 + 34$

7. $18 + 22$

8. $137 + 82$

9. $187 + 43$

10. $80 + 53$

11. $51 + 24$

12. $200 + 54$

13. $90 + 14$

14. $171 + 27$

15. $78 + 78$

16. $38 + 87$

17. $86 + 51$

18. $46 + 13$

19. $141 + 39$

20. $36 + 41$

1.   167
  + 19

2.   145
  + 41

3.   41
  + 24

4.   182
  + 53

5.   192
  + 61

6.   50
  + 56

7.   64
  + 88

8.   168
  + 60

9.   72
  + 33

10.   72
  + 17

11.   88
  + 68

12.   116
  + 20

13.   17
  + 71

14.   67
  + 16

15.   166
  + 83

16.   183
  + 33

17.   165
  + 68

18.   116
  + 58

19.   177
  + 46

20.   116
  + 14

1. $\begin{array}{r} 176 \\ + 76 \\ \hline \end{array}$ 2. $\begin{array}{r} 146 \\ + 13 \\ \hline \end{array}$ 3. $\begin{array}{r} 148 \\ + 84 \\ \hline \end{array}$ 4. $\begin{array}{r} 154 \\ + 54 \\ \hline \end{array}$

5. $\begin{array}{r} 90 \\ + 29 \\ \hline \end{array}$ 6. $\begin{array}{r} 152 \\ + 80 \\ \hline \end{array}$ 7. $\begin{array}{r} 148 \\ + 32 \\ \hline \end{array}$ 8. $\begin{array}{r} 134 \\ + 34 \\ \hline \end{array}$

9. $\begin{array}{r} 118 \\ + 53 \\ \hline \end{array}$ 10. $\begin{array}{r} 28 \\ + 74 \\ \hline \end{array}$ 11. $\begin{array}{r} 135 \\ + 60 \\ \hline \end{array}$ 12. $\begin{array}{r} 121 \\ + 71 \\ \hline \end{array}$

13. $\begin{array}{r} 194 \\ + 15 \\ \hline \end{array}$ 14. $\begin{array}{r} 48 \\ + 38 \\ \hline \end{array}$ 15. $\begin{array}{r} 149 \\ + 47 \\ \hline \end{array}$ 16. $\begin{array}{r} 18 \\ + 33 \\ \hline \end{array}$

17. $\begin{array}{r} 29 \\ + 23 \\ \hline \end{array}$ 18. $\begin{array}{r} 43 \\ + 58 \\ \hline \end{array}$ 19. $\begin{array}{r} 85 \\ + 19 \\ \hline \end{array}$ 20. $\begin{array}{r} 66 \\ + 18 \\ \hline \end{array}$

1.   136
  + 54

2.   39
  + 43

3.   143
  + 99

4.   87
  + 125

5.   127
  + 54

6.   82
  + 24

7.   59
  + 124

8.   123
  + 87

9.   141
  + 162

10.   56
  + 171

11.   58
  + 67

12.   66
  + 157

13.   18
  + 92

14.   188
  + 175

15.   186
  + 19

16.   53
  + 187

17.   60
  + 182

18.   45
  + 23

19.   85
  + 111

20.   19
  + 71

1.   111
   + 112
   ....

2.   139
   + 118
   ....

3.   11
   + 94
   ....

4.   164
   + 77
   ....

5.   88
   + 110
   ....

6.   12
   + 41
   ....

7.   152
   + 185
   ....

8.   42
   + 26
   ....

9.   89
   + 63
   ....

10.   108
    + 55
    ....

11.   155
    + 15
    ....

12.   64
    + 129
    ....

13.   88
    + 43
    ....

14.   83
    + 24
    ....

15.   198
    + 166
    ....

16.   35
    + 12
    ....

17.   86
    + 158
    ....

18.   160
    + 37
    ....

19.   102
    + 69
    ....

20.   12
    + 165
    ....

1.　　21
　+ 150
　..................

2.　　79
　+ 122
　..................

3.　　85
　+ 200
　..................

4.　154
　+ 195
　..................

5.　　35
　+ 15
　..................

6.　124
　+　30
　..................

7.　　62
　+ 143
　..................

8.　　55
　+ 96
　..................

9.　111
　+　53
　..................

10.　　73
　+ 188
　..................

11.　175
　+ 154
　..................

12.　　21
　+ 40
　..................

13.　176
　+ 100
　..................

14.　　38
　+ 106
　..................

15.　　61
　+ 64
　..................

16.　179
　+　92
　..................

17.　　39
　+ 43
　..................

18.　154
　+　60
　..................

19.　190
　+ 182
　..................

20.　136
　+　66
　..................

1. 172
  + 58
  ......

2. 144
  + 171
  ......

3. 36
  + 42
  ......

4. 191
  + 28
  ......

5. 189
  + 195
  ......

6. 107
  + 129
  ......

7. 117
  + 18
  ......

8. 52
  + 112
  ......

9. 59
  + 23
  ......

10. 187
  + 15
  ......

11. 26
  + 196
  ......

12. 171
  + 51
  ......

13. 69
  + 175
  ......

14. 145
  + 187
  ......

15. 50
  + 96
  ......

16. 121
  + 151
  ......

17. 45
  + 151
  ......

18. 94
  + 98
  ......

19. 129
  + 134
  ......

20. 172
  + 22
  ......

1. 97
+ 14

2. 178
+ 104

3. 84
+ 156

4. 94
+ 166

5. 124
+ 154

6. 182
+ 125

7. 50
+ 175

8. 36
+ 161

9. 154
+ 40

10. 137
+ 193

11. 88
+ 15

12. 187
+ 158

13. 87
+ 194

14. 98
+ 104

15. 182
+ 118

16. 20
+ 36

17. 32
+ 122

18. 125
+ 176

19. 87
+ 149

20. 66
+ 158

1.   68
  + 191
_____

2.   184
  + 16
_____

3.   144
  + 156
_____

4.   103
  + 128
_____

5.   60
  + 60
_____

6.   85
  + 15
_____

7.   174
  + 179
_____

8.   22
  + 156
_____

9.   190
  + 140
_____

10.   59
  + 193
_____

11.   117
  + 21
_____

12.   188
  + 175
_____

13.   96
  + 162
_____

14.   70
  + 114
_____

15.   149
  + 42
_____

16.   159
  + 67
_____

17.   132
  + 176
_____

18.   100
  + 191
_____

19.   148
  + 19
_____

20.   113
  + 154
_____

SUBTRACTION

SUBTRACTION

1.   6
   - 3

2.   75
   - 12

3.   51
   - 12

4.   51
   - 22

5.   15
   - 11

6.   58
   - 9

7.   61
   - 29

8.   22
   - 1

9.   28
   - 17

10.   50
   - 29

11.   91
   - 21

12.   34
   - 4

13.   82
   - 27

14.   60
   - 21

15.   30
   - 21

16.   54
   - 23

17.   41
   - 1

18.   68
   - 23

19.   86
   - 8

20.   30
   - 15

1.   44
   - 14
   .....

2.   16
   - 13
   .....

3.   86
   - 19
   .....

4.   30
   - 14
   .....

5.   74
   - 20
   .....

6.   74
   - 9
   .....

7.   67
   - 13
   .....

8.   38
   - 8
   .....

9.   74
   - 18
   .....

10.   91
   - 17
   .....

11.   35
   - 11
   .....

12.   96
   - 9
   .....

13.   1
   - 1
   .....

14.   62
   - 10
   .....

15.   71
   - 19
   .....

16.   5
   - 2
   .....

17.   47
   - 15
   .....

18.   6
   - 5
   .....

19.   92
   - 17
   .....

20.   38
   - 11
   .....

1. 98
    - 30
    ....

2. 73
    - 5
    ....

3. 63
    - 16
    ....

4. 78
    - 7
    ....

5. 38
    - 17
    ....

6. 32
    - 7
    ....

7. 21
    - 12
    ....

8. 80
    - 14
    ....

9. 17
    - 16
    ....

10. 73
    - 15
    ....

11. 83
    - 20
    ....

12. 32
    - 27
    ....

13. 71
    - 3
    ....

14. 52
    - 9
    ....

15. 3
    - 3
    ....

16. 35
    - 2
    ....

17. 72
    - 9
    ....

18. 38
    - 23
    ....

19. 18
    - 2
    ....

20. 42
    - 2
    ....

1. 62
    - 23

2. 54
    - 22

3. 70
    - 19

4. 97
    - 18

5. 62
    - 30

6. 8
    - 3

7. 92
    - 32

8. 23
    - 12

9. 80
    - 32

10. 8
    - 4

11. 29
    - 29

12. 32
    - 26

13. 70
    - 28

14. 24
    - 17

15. 11
    - 7

16. 69
    - 36

17. 33
    - 10

18. 94
    - 41

19. 2
    - 2

20. 33
    - 8

1. 15
   - 6
   ----

2. 25
   - 5
   ----

3. 89
   - 19
   ----

4. 89
   - 33
   ----

5. 68
   - 21
   ----

6. 49
   - 18
   ----

7. 31
   - 28
   ----

8. 59
   - 27
   ----

9. 13
   - 8
   ----

10. 48
    - 34
    ----

11. 12
    - 3
    ----

12. 5
    - 5
    ----

13. 99
    - 38
    ----

14. 23
    - 13
    ----

15. 88
    - 17
    ----

16. 69
    - 14
    ----

17. 76
    - 10
    ----

18. 6
    - 4
    ----

19. 53
    - 38
    ----

20. 87
    - 48
    ----

1.   16
     - 7

2.   76
     - 20

3.   72
     - 13

4.   57
     - 8

5.   84
     - 7

6.   99
     - 19

7.   67
     - 27

8.   58
     - 17

9.   59
     - 40

10.   27
     - 23

11.   14
     - 9

12.   34
     - 2

13.   78
     - 7

14.   98
     - 38

15.   14
     - 14

16.   23
     - 7

17.   64
     - 37

18.   80
     - 38

19.   5
     - 2

20.   53
     - 23

1. $\begin{array}{r} 26 \\ -\ 22 \\ \hline \end{array}$
2. $\begin{array}{r} 99 \\ -\ 44 \\ \hline \end{array}$
3. $\begin{array}{r} 89 \\ -\ 39 \\ \hline \end{array}$
4. $\begin{array}{r} 72 \\ -\ 47 \\ \hline \end{array}$

5. $\begin{array}{r} 72 \\ -\ 42 \\ \hline \end{array}$
6. $\begin{array}{r} 29 \\ -\ 15 \\ \hline \end{array}$
7. $\begin{array}{r} 28 \\ -\ 23 \\ \hline \end{array}$
8. $\begin{array}{r} 23 \\ -\ 15 \\ \hline \end{array}$

9. $\begin{array}{r} 86 \\ -\ 13 \\ \hline \end{array}$
10. $\begin{array}{r} 15 \\ -\ 10 \\ \hline \end{array}$
11. $\begin{array}{r} 21 \\ -\ 7 \\ \hline \end{array}$
12. $\begin{array}{r} 32 \\ -\ 14 \\ \hline \end{array}$

13. $\begin{array}{r} 49 \\ -\ 37 \\ \hline \end{array}$
14. $\begin{array}{r} 18 \\ -\ 8 \\ \hline \end{array}$
15. $\begin{array}{r} 72 \\ -\ 16 \\ \hline \end{array}$
16. $\begin{array}{r} 69 \\ -\ 15 \\ \hline \end{array}$

17. $\begin{array}{r} 51 \\ -\ 46 \\ \hline \end{array}$
18. $\begin{array}{r} 21 \\ -\ 17 \\ \hline \end{array}$
19. $\begin{array}{r} 51 \\ -\ 44 \\ \hline \end{array}$
20. $\begin{array}{r} 21 \\ -\ 4 \\ \hline \end{array}$

1.   61
  - 20

2.   22
  - 11

3.   16
  - 2

4.   98
  - 19

5.   62
  - 23

6.   13
  - 13

7.   6
  - 2

8.   16
  - 5

9.   25
  - 18

10.   93
  - 25

11.   54
  - 47

12.   28
  - 11

13.   63
  - 44

14.   64
  - 27

15.   50
  - 12

16.   30
  - 27

17.   49
  - 4

18.   50
  - 6

19.   76
  - 30

20.   46
  - 22

1. $38 - 19$

2. $86 - 21$

3. $16 - 5$

4. $50 - 50$

5. $54 - 5$

6. $46 - 31$

7. $28 - 15$

8. $49 - 34$

9. $33 - 21$

10. $7 - 7$

11. $12 - 2$

12. $68 - 16$

13. $46 - 24$

14. $24 - 15$

15. $42 - 35$

16. $97 - 45$

17. $77 - 11$

18. $10 - 6$

19. $97 - 12$

20. $51 - 5$

1.   98
    - 46
    ....

2.   58
    - 37
    ....

3.   10
    - 7
    ....

4.   88
    - 6
    ....

5.   41
    - 39
    ....

6.   70
    - 7
    ....

7.   14
    - 14
    ....

8.   77
    - 18
    ....

9.   97
    - 45
    ....

10. 54
    - 19
    ....

11. 97
    - 4
    ....

12. 17
    - 9
    ....

13. 47
    - 25
    ....

14. 7
    - 7
    ....

15. 5
    - 4
    ....

16. 74
    - 21
    ....

17. 27
    - 24
    ....

18. 10
    - 4
    ....

19. 21
    - 8
    ....

20. 69
    - 11
    ....

1. $140 - 11$

2. $8 - 7$

3. $7 - 2$

4. $86 - 7$

5. $101 - 44$

6. $25 - 5$

7. $63 - 17$

8. $124 - 5$

9. $39 - 12$

10. $75 - 35$

11. $27 - 19$

12. $30 - 5$

13. $66 - 21$

14. $39 - 4$

15. $106 - 20$

16. $17 - 10$

17. $131 - 45$

18. $149 - 12$

19. $9 - 3$

20. $81 - 35$

1.   117
  - 27
----

2.   60
  - 16
----

3.   117
  - 32
----

4.   39
  - 21
----

5.   50
  - 32
----

6.   5
  - 2
----

7.   109
  - 41
----

8.   85
  - 30
----

9.   98
  - 31
----

10.   24
  - 21
----

11.   108
  - 36
----

12.   127
  - 19
----

13.   49
  - 12
----

14.   29
  - 11
----

15.   107
  - 37
----

16.   33
  - 6
----

17.   128
  - 45
----

18.   135
  - 32
----

19.   30
  - 22
----

20.   122
  - 18
----

1.   67
   - 47

2.   120
   - 30

3.   63
   - 46

4.   34
   - 29

5.   138
   - 47

6.   61
   - 46

7.   119
   - 41

8.   81
   - 28

9.   78
   - 15

10.   8
   - 4

11.   9
   - 6

12.   49
   - 7

13.   20
   - 17

14.   50
   - 20

15.   145
   - 42

16.   143
   - 10

17.   114
   - 14

18.   135
   - 21

19.   57
   - 32

20.   28
   - 13

1.   88
   - 42
   ----

2.   131
   -   6
   ----

3.   147
   - 13
   ----

4.   27
   - 16
   ----

5.   137
   -   9
   ----

6.   29
   - 26
   ----

7.   15
   - 10
   ----

8.   85
   - 37
   ----

9.   4
   - 3
   ----

10.   134
    - 23
    ----

11.   30
    - 12
    ----

12.   67
    - 30
    ----

13.   45
    - 26
    ----

14.   27
    - 5
    ----

15.   109
    -   7
    ----

16.   117
    - 34
    ----

17.   24
    - 15
    ----

18.   126
    - 47
    ----

19.   70
    - 28
    ----

20.   14
    - 14
    ----

1. 13
    - 3

2. 124
    - 41

3. 56
    - 27

4. 113
    - 27

5. 119
    - 32

6. 41
    - 20

7. 27
    - 25

8. 137
    - 8

9. 45
    - 30

10. 61
    - 32

11. 35
    - 28

12. 8
    - 8

13. 48
    - 21

14. 130
    - 7

15. 26
    - 17

16. 80
    - 11

17. 115
    - 33

18. 136
    - 25

19. 38
    - 11

20. 65
    - 35

1. 98
    - 33
    ....

2. 55
    - 11
    ....

3. 18
    - 3
    ....

4. 105
    - 27
    ....

5. 4
    - 3
    ....

6. 132
    - 4
    ....

7. 33
    - 24
    ....

8. 109
    - 22
    ....

9. 35
    - 22
    ....

10. 60
    - 49
    ....

11. 34
    - 33
    ....

12. 148
    - 44
    ....

13. 102
    - 39
    ....

14. 46
    - 29
    ....

15. 68
    - 32
    ....

16. 71
    - 19
    ....

17. 112
    - 25
    ....

18. 7
    - 3
    ....

19. 69
    - 21
    ....

20. 52
    - 23
    ....

1. $\begin{array}{r} 119 \\ -\ \ 6 \\ \hline \end{array}$ 2. $\begin{array}{r} 13 \\ -\ 6 \\ \hline \end{array}$ 3. $\begin{array}{r} 146 \\ -\ 27 \\ \hline \end{array}$ 4. $\begin{array}{r} 46 \\ -\ 35 \\ \hline \end{array}$

5. $\begin{array}{r} 104 \\ -\ 24 \\ \hline \end{array}$ 6. $\begin{array}{r} 145 \\ -\ 42 \\ \hline \end{array}$ 7. $\begin{array}{r} 40 \\ -\ 38 \\ \hline \end{array}$ 8. $\begin{array}{r} 55 \\ -\ 28 \\ \hline \end{array}$

9. $\begin{array}{r} 15 \\ -\ 6 \\ \hline \end{array}$ 10. $\begin{array}{r} 70 \\ -\ 40 \\ \hline \end{array}$ 11. $\begin{array}{r} 34 \\ -\ 6 \\ \hline \end{array}$ 12. $\begin{array}{r} 29 \\ -\ 20 \\ \hline \end{array}$

13. $\begin{array}{r} 89 \\ -\ 4 \\ \hline \end{array}$ 14. $\begin{array}{r} 127 \\ -\ 13 \\ \hline \end{array}$ 15. $\begin{array}{r} 110 \\ -\ 19 \\ \hline \end{array}$ 16. $\begin{array}{r} 94 \\ -\ 48 \\ \hline \end{array}$

17. $\begin{array}{r} 135 \\ -\ 50 \\ \hline \end{array}$ 18. $\begin{array}{r} 99 \\ -\ 28 \\ \hline \end{array}$ 19. $\begin{array}{r} 92 \\ -\ 26 \\ \hline \end{array}$ 20. $\begin{array}{r} 108 \\ -\ 44 \\ \hline \end{array}$

1. $103 - 5$

2. $4 - 3$

3. $125 - 2$

4. $139 - 14$

5. $45 - 37$

6. $139 - 30$

7. $88 - 30$

8. $32 - 14$

9. $101 - 12$

10. $126 - 19$

11. $134 - 16$

12. $12 - 7$

13. $116 - 15$

14. $142 - 11$

15. $131 - 2$

16. $96 - 11$

17. $106 - 40$

18. $31 - 17$

19. $120 - 9$

20. $23 - 19$

1. $\begin{array}{r} 10 \\ -\ 8 \\ \hline \end{array}$ 2. $\begin{array}{r} 59 \\ -16 \\ \hline \end{array}$ 3. $\begin{array}{r} 28 \\ -21 \\ \hline \end{array}$ 4. $\begin{array}{r} 134 \\ -\ 39 \\ \hline \end{array}$

5. $\begin{array}{r} 60 \\ -35 \\ \hline \end{array}$ 6. $\begin{array}{r} 92 \\ -27 \\ \hline \end{array}$ 7. $\begin{array}{r} 12 \\ -\ 2 \\ \hline \end{array}$ 8. $\begin{array}{r} 12 \\ -12 \\ \hline \end{array}$

9. $\begin{array}{r} 7 \\ -6 \\ \hline \end{array}$ 10. $\begin{array}{r} 37 \\ -29 \\ \hline \end{array}$ 11. $\begin{array}{r} 76 \\ -26 \\ \hline \end{array}$ 12. $\begin{array}{r} 103 \\ -\ 9 \\ \hline \end{array}$

13. $\begin{array}{r} 6 \\ -6 \\ \hline \end{array}$ 14. $\begin{array}{r} 94 \\ -15 \\ \hline \end{array}$ 15. $\begin{array}{r} 117 \\ -\ 41 \\ \hline \end{array}$ 16. $\begin{array}{r} 78 \\ -16 \\ \hline \end{array}$

17. $\begin{array}{r} 88 \\ -30 \\ \hline \end{array}$ 18. $\begin{array}{r} 115 \\ -\ 29 \\ \hline \end{array}$ 19. $\begin{array}{r} 78 \\ -49 \\ \hline \end{array}$ 20. $\begin{array}{r} 28 \\ -16 \\ \hline \end{array}$

1.   19
    - 13
	2. 2
	    - 2
		3. 27
		    - 10
			4. 142
			    - 25

5.   41
    - 22
	6. 72
	    - 5
		7. 83
		    - 35
			8. 101
			    - 26

9.   70
    - 43
	10. 7
	    - 5
		11. 49
		    - 5
			12. 136
			    - 37

13.   69
    - 13
	14. 8
	    - 7
		15. 80
		    - 12
			16. 77
			    - 15

17.   108
    - 47
	18. 66
	    - 25
		19. 149
		    - 24
			20. 98
			    - 41

ANSWERS

ANSWERS

1. $5 + 25 = 30$
2. $14 + 4 = 18$
3. $14 + 21 = 35$
4. $18 + 33 = 51$

5. $23 + 25 = 48$
6. $35 + 17 = 52$
7. $7 + 6 = 13$
8. $8 + 28 = 36$

9. $11 + 34 = 45$
10. $14 + 36 = 50$
11. $7 + 20 = 27$
12. $31 + 6 = 37$

13. $33 + 37 = 70$
14. $10 + 3 = 13$
15. $36 + 26 = 62$
16. $19 + 38 = 57$

17. $20 + 40 = 60$
18. $28 + 21 = 49$
19. $18 + 7 = 25$
20. $12 + 12 = 24$

1. $12 + 16 = 28$
2. $37 + 37 = 74$
3. $6 + 37 = 43$
4. $38 + 7 = 45$

5. $19 + 27 = 46$
6. $11 + 5 = 16$
7. $22 + 3 = 25$
8. $9 + 23 = 32$

9. $15 + 33 = 48$
10. $17 + 33 = 50$
11. $29 + 26 = 55$
12. $14 + 36 = 50$

13. $20 + 10 = 30$
14. $32 + 28 = 60$
15. $1 + 4 = 5$
16. $27 + 10 = 37$

17. $36 + 35 = 71$
18. $1 + 13 = 14$
19. $27 + 9 = 36$
20. $33 + 31 = 64$

1. $6 + 11 = 17$
2. $7 + 16 = 23$
3. $40 + 2 = 42$
4. $9 + 28 = 37$

5. $3 + 37 = 40$
6. $35 + 7 = 42$
7. $1 + 26 = 27$
8. $30 + 27 = 57$

9. $21 + 12 = 33$
10. $7 + 18 = 25$
11. $22 + 29 = 51$
12. $6 + 24 = 30$

13. $33 + 26 = 59$
14. $18 + 19 = 37$
15. $7 + 15 = 22$
16. $10 + 26 = 36$

17. $31 + 36 = 67$
18. $26 + 33 = 59$
19. $36 + 30 = 66$
20. $3 + 29 = 32$

1. $24 + 29 = 53$
2. $21 + 28 = 49$
3. $27 + 10 = 37$
4. $16 + 5 = 21$

5. $40 + 37 = 77$
6. $31 + 32 = 63$
7. $9 + 18 = 27$
8. $24 + 22 = 46$

9. $28 + 19 = 47$
10. $2 + 29 = 31$
11. $25 + 12 = 37$
12. $31 + 6 = 37$

13. $10 + 19 = 29$
14. $31 + 30 = 61$
15. $22 + 10 = 32$
16. $28 + 14 = 42$

17. $7 + 28 = 35$
18. $6 + 6 = 12$
19. $20 + 14 = 34$
20. $40 + 13 = 53$

1. 43 + 14 **57**	2. 39 + 19 **58**	3. 21 + 39 **60**	4. 36 + 28 **64**
5. 60 + 37 **97**	6. 26 + 30 **56**	7. 31 + 2 **33**	8. 13 + 33 **46**
9. 13 + 15 **28**	10. 65 + 8 **73**	11. 58 + 2 **60**	12. 30 + 37 **67**
13. 46 + 8 **54**	14. 39 + 12 **51**	15. 33 + 30 **63**	16. 54 + 19 **73**
17. 12 + 14 **26**	18. 16 + 16 **32**	19. 48 + 32 **80**	20. 58 + 20 **78**

1. 17 + 5 **22**	2. 12 + 36 **48**	3. 41 + 5 **46**	4. 57 + 32 **89**
5. 23 + 20 **43**	6. 47 + 6 **53**	7. 25 + 14 **39**	8. 6 + 19 **25**
9. 18 + 10 **28**	10. 44 + 3 **47**	11. 61 + 39 **100**	12. 37 + 5 **42**
13. 67 + 26 **93**	14. 33 + 36 **69**	15. 61 + 23 **84**	16. 28 + 24 **52**
17. 19 + 33 **52**	18. 56 + 31 **87**	19. 60 + 18 **78**	20. 49 + 24 **73**

1. 25 + 18 **43**	2. 44 + 4 **48**	3. 51 + 30 **81**	4. 6 + 11 **17**
5. 47 + 19 **66**	6. 26 + 8 **34**	7. 26 + 12 **38**	8. 55 + 15 **70**
9. 35 + 29 **64**	10. 29 + 37 **66**	11. 9 + 2 **11**	12. 36 + 12 **48**
13. 7 + 3 **10**	14. 17 + 40 **57**	15. 37 + 18 **55**	16. 28 + 17 **45**
17. 22 + 21 **43**	18. 19 + 30 **49**	19. 57 + 9 **66**	20. 7 + 38 **45**

1. 36 + 4 **40**	2. 51 + 10 **61**	3. 7 + 39 **46**	4. 62 + 34 **96**
5. 2 + 2 **4**	6. 38 + 12 **50**	7. 44 + 18 **62**	8. 67 + 34 **101**
9. 18 + 24 **42**	10. 4 + 31 **35**	11. 4 + 7 **11**	12. 36 + 23 **59**
13. 42 + 29 **71**	14. 24 + 4 **28**	15. 33 + 37 **70**	16. 58 + 19 **77**
17. 16 + 16 **32**	18. 32 + 35 **67**	19. 38 + 22 **60**	20. 58 + 12 **70**

1. 42 + 44 86	2. 44 + 34 78	3. 15 + 36 51	4. 33 + 27 60
5. 38 + 11 49	6. 39 + 36 75	7. 20 + 35 55	8. 60 + 68 128
9. 19 + 88 107	10. 28 + 60 88	11. 18 + 23 41	12. 22 + 79 101
13. 66 + 24 90	14. 66 + 85 151	15. 59 + 4 63	16. 37 + 78 115
17. 47 + 45 92	18. 36 + 14 50	19. 49 + 7 56	20. 14 + 26 40

1. 38 + 41 79	2. 39 + 39 78	3. 50 + 55 105	4. 57 + 60 117
5. 66 + 62 128	6. 17 + 8 25	7. 65 + 3 68	8. 64 + 68 132
9. 44 + 19 63	10. 39 + 47 86	11. 12 + 51 63	12. 69 + 76 145
13. 60 + 41 101	14. 51 + 17 68	15. 51 + 15 66	16. 61 + 85 146
17. 47 + 22 69	18. 59 + 29 88	19. 64 + 24 88	20. 30 + 71 101

1. 199 + 76 275	2. 98 + 43 141	3. 157 + 68 225	4. 39 + 79 118
5. 16 + 26 42	6. 161 + 78 239	7. 116 + 77 193	8. 60 + 20 80
9. 116 + 46 162	10. 161 + 76 237	11. 115 + 30 145	12. 13 + 45 58
13. 116 + 80 196	14. 61 + 12 73	15. 183 + 15 198	16. 150 + 19 169
17. 112 + 49 161	18. 123 + 86 209	19. 198 + 30 228	20. 88 + 61 149

1. 96 + 20 116	2. 192 + 87 279	3. 98 + 64 162	4. 71 + 15 86
5. 56 + 35 91	6. 36 + 34 70	7. 18 + 22 40	8. 137 + 82 219
9. 187 + 43 230	10. 80 + 53 133	11. 51 + 24 75	12. 200 + 54 254
13. 90 + 14 104	14. 171 + 27 198	15. 78 + 78 156	16. 38 + 87 125
17. 86 + 51 137	18. 46 + 13 59	19. 141 + 39 180	20. 36 + 41 77

1. 167
+ 19
186

2. 145
+ 41
186

3. 41
+ 24
65

4. 182
+ 53
235

5. 192
+ 61
253

6. 50
+ 56
106

7. 64
+ 88
152

8. 168
+ 60
228

9. 72
+ 33
105

10. 72
+ 17
89

11. 88
+ 68
156

12. 116
+ 20
136

13. 17
+ 71
88

14. 67
+ 16
83

15. 166
+ 83
249

16. 183
+ 33
216

17. 165
+ 68
233

18. 116
+ 58
174

19. 177
+ 46
223

20. 116
+ 14
130

1. 176
+ 76
252

2. 146
+ 13
159

3. 148
+ 84
232

4. 154
+ 54
208

5. 90
+ 29
119

6. 152
+ 80
232

7. 148
+ 32
180

8. 134
+ 34
168

9. 118
+ 53
171

10. 28
+ 74
102

11. 135
+ 60
195

12. 121
+ 71
192

13. 194
+ 15
209

14. 48
+ 38
86

15. 149
+ 47
196

16. 18
+ 33
51

17. 29
+ 23
52

18. 43
+ 58
101

19. 85
+ 19
104

20. 66
+ 18
84

1. 136
+ 54
190

2. 39
+ 43
82

3. 143
+ 99
242

4. 87
+ 125
212

5. 127
+ 54
181

6. 82
+ 24
106

7. 59
+ 124
183

8. 123
+ 87
210

9. 141
+ 162
303

10. 56
+ 171
227

11. 58
+ 67
125

12. 66
+ 157
223

13. 18
+ 92
110

14. 188
+ 175
363

15. 186
+ 19
205

16. 53
+ 187
240

17. 60
+ 182
242

18. 45
+ 23
68

19. 85
+ 111
196

20. 19
+ 71
90

1. 111
+ 112
223

2. 139
+ 118
257

3. 11
+ 94
105

4. 164
+ 77
241

5. 88
+ 110
198

6. 12
+ 41
53

7. 152
+ 185
337

8. 42
+ 26
68

9. 89
+ 63
152

10. 108
+ 55
163

11. 155
+ 15
170

12. 64
+ 129
193

13. 88
+ 43
131

14. 83
+ 24
107

15. 198
+ 166
364

16. 35
+ 12
47

17. 86
+ 158
244

18. 160
+ 37
197

19. 102
+ 69
171

20. 12
+ 165
177

1. 21 + 150 = 171
2. 79 + 122 = 201
3. 85 + 200 = 285
4. 154 + 195 = 349

5. 35 + 15 = 50
6. 124 + 30 = 154
7. 62 + 143 = 205
8. 55 + 96 = 151

9. 111 + 53 = 164
10. 73 + 188 = 261
11. 175 + 154 = 329
12. 21 + 40 = 61

13. 176 + 100 = 276
14. 38 + 106 = 144
15. 61 + 64 = 125
16. 179 + 92 = 271

17. 39 + 43 = 82
18. 154 + 60 = 214
19. 190 + 182 = 372
20. 136 + 66 = 202

1. 172 + 58 = 230
2. 144 + 171 = 315
3. 36 + 42 = 78
4. 191 + 28 = 219

5. 189 + 195 = 384
6. 107 + 129 = 236
7. 117 + 18 = 135
8. 52 + 112 = 164

9. 59 + 23 = 82
10. 187 + 15 = 202
11. 26 + 196 = 222
12. 171 + 51 = 222

13. 69 + 175 = 244
14. 145 + 187 = 332
15. 50 + 96 = 146
16. 121 + 151 = 272

17. 45 + 151 = 196
18. 94 + 98 = 192
19. 129 + 134 = 263
20. 172 + 22 = 194

1. 97 + 14 = 111
2. 178 + 104 = 282
3. 84 + 156 = 240
4. 94 + 166 = 260

5. 124 + 154 = 278
6. 182 + 125 = 307
7. 50 + 175 = 225
8. 36 + 161 = 197

9. 154 + 40 = 194
10. 137 + 193 = 330
11. 88 + 15 = 103
12. 187 + 158 = 345

13. 87 + 194 = 281
14. 98 + 104 = 202
15. 182 + 118 = 300
16. 20 + 36 = 56

17. 32 + 122 = 154
18. 125 + 176 = 301
19. 87 + 149 = 236
20. 66 + 158 = 224

1. 68 + 191 = 259
2. 184 + 16 = 200
3. 144 + 156 = 300
4. 103 + 128 = 231

5. 60 + 60 = 120
6. 85 + 15 = 100
7. 174 + 179 = 353
8. 22 + 156 = 178

9. 190 + 140 = 330
10. 59 + 193 = 252
11. 117 + 21 = 138
12. 188 + 175 = 363

13. 96 + 162 = 258
14. 70 + 114 = 184
15. 149 + 42 = 191
16. 159 + 67 = 226

17. 132 + 176 = 308
18. 100 + 191 = 291
19. 148 + 19 = 167
20. 113 + 154 = 267

1. 6
  - 3
  ___
    3

2. 75
  - 12
  ___
   63

3. 51
  - 12
  ___
   39

4. 51
  - 22
  ___
   29

5. 15
  - 11
  ___
    4

6. 58
  - 9
  ___
   49

7. 61
  - 29
  ___
   32

8. 22
  - 1
  ___
   21

9. 28
  - 17
  ___
   11

10. 50
  - 29
  ___
   21

11. 91
  - 21
  ___
   70

12. 34
  - 4
  ___
   30

13. 82
  - 27
  ___
   55

14. 60
  - 21
  ___
   39

15. 30
  - 21
  ___
    9

16. 54
  - 23
  ___
   31

17. 41
  - 1
  ___
   40

18. 68
  - 23
  ___
   45

19. 86
  - 8
  ___
   78

20. 30
  - 15
  ___
   15

1. 44
  - 14
  ___
   30

2. 16
  - 13
  ___
    3

3. 86
  - 19
  ___
   67

4. 30
  - 14
  ___
   16

5. 74
  - 20
  ___
   54

6. 74
  - 9
  ___
   65

7. 67
  - 13
  ___
   54

8. 38
  - 8
  ___
   30

9. 74
  - 18
  ___
   56

10. 91
  - 17
  ___
   74

11. 35
  - 11
  ___
   24

12. 96
  - 9
  ___
   87

13. 1
  - 1
  ___
    0

14. 62
  - 10
  ___
   52

15. 71
  - 19
  ___
   52

16. 5
  - 2
  ___
    3

17. 47
  - 15
  ___
   32

18. 6
  - 5
  ___
    1

19. 92
  - 17
  ___
   75

20. 38
  - 11
  ___
   27

1. 98
  - 30
  ___
   68

2. 73
  - 5
  ___
   68

3. 63
  - 16
  ___
   47

4. 78
  - 7
  ___
   71

5. 38
  - 17
  ___
   21

6. 32
  - 7
  ___
   25

7. 21
  - 12
  ___
    9

8. 80
  - 14
  ___
   66

9. 17
  - 16
  ___
    1

10. 73
  - 15
  ___
   58

11. 83
  - 20
  ___
   63

12. 32
  - 27
  ___
    5

13. 71
  - 3
  ___
   68

14. 52
  - 9
  ___
   43

15. 3
  - 3
  ___
    0

16. 35
  - 2
  ___
   33

17. 72
  - 9
  ___
   63

18. 38
  - 23
  ___
   15

19. 18
  - 2
  ___
   16

20. 42
  - 2
  ___
   40

1. 62
  - 23
  ___
   39

2. 54
  - 22
  ___
   32

3. 70
  - 19
  ___
   51

4. 97
  - 18
  ___
   79

5. 62
  - 30
  ___
   32

6.   8
  - 3
  ___
    5

7. 92
  - 32
  ___
   60

8. 23
  - 12
  ___
   11

9. 80
  - 32
  ___
   48

10. 8
  - 4
  ___
    4

11. 29
  - 29
  ___
    0

12. 32
  - 26
  ___
    6

13. 70
  - 28
  ___
   42

14. 24
  - 17
  ___
    7

15. 11
  - 7
  ___
    4

16. 69
  - 36
  ___
   33

17. 33
  - 10
  ___
   23

18. 94
  - 41
  ___
   53

19. 2
  - 2
  ___
    0

20. 33
  - 8
  ___
   25

1. 15
   - 6
   9

2. 25
   - 5
   20

3. 89
   - 19
   70

4. 89
   - 33
   56

5. 68
   - 21
   47

6. 49
   - 18
   31

7. 31
   - 28
   3

8. 59
   - 27
   32

9. 13
   - 8
   5

10. 48
    - 34
    14

11. 12
    - 3
    9

12. 5
    - 5
    0

13. 99
    - 38
    61

14. 23
    - 13
    10

15. 88
    - 17
    71

16. 69
    - 14
    55

17. 76
    - 10
    66

18. 6
    - 4
    2

19. 53
    - 38
    15

20. 87
    - 48
    39

1. 16
   - 7
   9

2. 76
   - 20
   56

3. 72
   - 13
   59

4. 57
   - 8
   49

5. 84
   - 7
   77

6. 99
   - 19
   80

7. 67
   - 27
   40

8. 58
   - 17
   41

9. 59
   - 40
   19

10. 27
    - 23
    4

11. 14
    - 9
    5

12. 34
    - 2
    32

13. 78
    - 7
    71

14. 98
    - 38
    60

15. 14
    - 14
    0

16. 23
    - 7
    16

17. 64
    - 37
    27

18. 80
    - 38
    42

19. 5
    - 2
    3

20. 53
    - 23
    30

1. 26
   - 22
   4

2. 99
   - 44
   55

3. 89
   - 39
   50

4. 72
   - 47
   25

5. 72
   - 42
   30

6. 29
   - 15
   14

7. 28
   - 23
   5

8. 23
   - 15
   8

9. 86
   - 13
   73

10. 15
    - 10
    5

11. 21
    - 7
    14

12. 32
    - 14
    18

13. 49
    - 37
    12

14. 18
    - 8
    10

15. 72
    - 16
    56

16. 69
    - 15
    54

17. 51
    - 46
    5

18. 21
    - 17
    4

19. 51
    - 44
    7

20. 21
    - 4
    17

1. 61
   - 20
   41

2. 22
   - 11
   11

3. 16
   - 2
   14

4. 98
   - 19
   79

5. 62
   - 23
   39

6. 13
   - 13
   0

7. 6
   - 2
   4

8. 16
   - 5
   11

9. 25
   - 18
   7

10. 93
    - 25
    68

11. 54
    - 47
    7

12. 28
    - 11
    17

13. 63
    - 44
    19

14. 64
    - 27
    37

15. 50
    - 12
    38

16. 30
    - 27
    3

17. 49
    - 4
    45

18. 50
    - 6
    44

19. 76
    - 30
    46

20. 46
    - 22
    24

1. 38 − 19 = 19
2. 86 − 21 = 65
3. 16 − 5 = 11
4. 50 − 50 = 0

5. 54 − 5 = 49
6. 46 − 31 = 15
7. 28 − 15 = 13
8. 49 − 34 = 15

9. 33 − 21 = 12
10. 7 − 7 = 0
11. 12 − 2 = 10
12. 68 − 16 = 52

13. 46 − 24 = 22
14. 24 − 15 = 9
15. 42 − 35 = 7
16. 97 − 45 = 52

17. 77 − 11 = 66
18. 10 − 6 = 4
19. 97 − 12 = 85
20. 51 − 5 = 46

1. 98 − 46 = 52
2. 58 − 37 = 21
3. 10 − 7 = 3
4. 88 − 6 = 82

5. 41 − 39 = 2
6. 70 − 7 = 63
7. 14 − 14 = 0
8. 77 − 18 = 59

9. 97 − 45 = 52
10. 54 − 19 = 35
11. 97 − 4 = 93
12. 17 − 9 = 8

13. 47 − 25 = 22
14. 7 − 7 = 0
15. 5 − 4 = 1
16. 74 − 21 = 53

17. 27 − 24 = 3
18. 10 − 4 = 6
19. 21 − 8 = 13
20. 69 − 11 = 58

1. 140 − 11 = 129
2. 8 − 7 = 1
3. 7 − 2 = 5
4. 86 − 7 = 79

5. 101 − 44 = 57
6. 25 − 5 = 20
7. 63 − 17 = 46
8. 124 − 5 = 119

9. 39 − 12 = 27
10. 75 − 35 = 40
11. 27 − 19 = 8
12. 30 − 5 = 25

13. 66 − 21 = 45
14. 39 − 4 = 35
15. 106 − 20 = 86
16. 17 − 10 = 7

17. 131 − 45 = 86
18. 149 − 12 = 137
19. 9 − 3 = 6
20. 81 − 35 = 46

1. 117 − 27 = 90
2. 60 − 16 = 44
3. 117 − 32 = 85
4. 39 − 21 = 18

5. 50 − 32 = 18
6. 5 − 2 = 3
7. 109 − 41 = 68
8. 85 − 30 = 55

9. 98 − 31 = 67
10. 24 − 21 = 3
11. 108 − 36 = 72
12. 127 − 19 = 108

13. 49 − 12 = 37
14. 29 − 11 = 18
15. 107 − 37 = 70
16. 33 − 6 = 27

17. 128 − 45 = 83
18. 135 − 32 = 103
19. 30 − 22 = 8
20. 122 − 18 = 104

1. 67
− 47
20

2. 120
− 30
90

3. 63
− 46
17

4. 34
− 29
5

5. 138
− 47
91

6. 61
− 46
15

7. 119
− 41
78

8. 81
− 28
53

9. 78
− 15
63

10. 8
− 4
4

11. 9
− 6
3

12. 49
− 7
42

13. 20
− 17
3

14. 50
− 20
30

15. 145
− 42
103

16. 143
− 10
133

17. 114
− 14
100

18. 135
− 21
114

19. 57
− 32
25

20. 28
− 13
15

1. 88
− 42
46

2. 131
− 6
125

3. 147
− 13
134

4. 27
− 16
11

5. 137
− 9
128

6. 29
− 26
3

7. 15
− 10
5

8. 85
− 37
48

9. 4
− 3
1

10. 134
− 23
111

11. 30
− 12
18

12. 67
− 30
37

13. 45
− 26
19

14. 27
− 5
22

15. 109
− 7
102

16. 117
− 34
83

17. 24
− 15
9

18. 126
− 47
79

19. 70
− 28
42

20. 14
− 14
0

1. 13
− 3
10

2. 124
− 41
83

3. 56
− 27
29

4. 113
− 27
86

5. 119
− 32
87

6. 41
− 20
21

7. 27
− 25
2

8. 137
− 8
129

9. 45
− 30
15

10. 61
− 32
29

11. 35
− 28
7

12. 8
− 8
0

13. 48
− 21
27

14. 130
− 7
123

15. 26
− 17
9

16. 80
− 11
69

17. 115
− 33
82

18. 136
− 25
111

19. 38
− 11
27

20. 65
− 35
30

1. 98
− 33
65

2. 55
− 11
44

3. 18
− 3
15

4. 105
− 27
78

5. 4
− 3
1

6. 132
− 4
128

7. 33
− 24
9

8. 109
− 22
87

9. 35
− 22
13

10. 60
− 49
11

11. 34
− 33
1

12. 148
− 44
104

13. 102
− 39
63

14. 46
− 29
17

15. 68
− 32
36

16. 71
− 19
52

17. 112
− 25
87

18. 7
− 3
4

19. 69
− 21
48

20. 52
− 23
29

1. 119
 - 6
 113

2. 13
 - 6
 7

3. 146
 - 27
 119

4. 46
 - 35
 11

5. 104
 - 24
 80

6. 145
 - 42
 103

7. 40
 - 38
 2

8. 55
 - 28
 27

9. 15
 - 6
 9

10. 70
 - 40
 30

11. 34
 - 6
 28

12. 29
 - 20
 9

13. 89
 - 4
 85

14. 127
 - 13
 114

15. 110
 - 19
 91

16. 94
 - 48
 46

17. 135
 - 50
 85

18. 99
 - 28
 71

19. 92
 - 26
 66

20. 108
 - 44
 64

1. 103
 - 5
 98

2. 4
 - 3
 1

3. 125
 - 2
 123

4. 139
 - 14
 125

5. 45
 - 37
 8

6. 139
 - 30
 109

7. 88
 - 30
 58

8. 32
 - 14
 18

9. 101
 - 12
 89

10. 126
 - 19
 107

11. 134
 - 16
 118

12. 12
 - 7
 5

13. 116
 - 15
 101

14. 142
 - 11
 131

15. 131
 - 2
 129

16. 96
 - 11
 85

17. 106
 - 40
 66

18. 31
 - 17
 14

19. 120
 - 9
 111

20. 23
 - 19
 4

1. 10
 - 8
 2

2. 59
 - 16
 43

3. 28
 - 21
 7

4. 134
 - 39
 95

5. 60
 - 35
 25

6. 92
 - 27
 65

7. 12
 - 2
 10

8. 12
 - 12
 0

9. 7
 - 6
 1

10. 37
 - 29
 8

11. 76
 - 26
 50

12. 103
 - 9
 94

13. 6
 - 6
 0

14. 94
 - 15
 79

15. 117
 - 41
 76

16. 78
 - 16
 62

17. 88
 - 30
 58

18. 115
 - 29
 86

19. 78
 - 49
 29

20. 28
 - 16
 12

1. 19
 - 13
 6

2. 2
 - 2
 0

3. 27
 - 10
 17

4. 142
 - 25
 117

5. 41
 - 22
 19

6. 72
 - 5
 67

7. 83
 - 35
 48

8. 101
 - 26
 75

9. 70
 - 43
 27

10. 7
 - 5
 2

11. 49
 - 5
 44

12. 136
 - 37
 99

13. 69
 - 13
 56

14. 8
 - 7
 1

15. 80
 - 12
 68

16. 77
 - 15
 62

17. 108
 - 47
 61

18. 66
 - 25
 41

19. 149
 - 24
 125

20. 98
 - 41
 57

Made in the USA
Las Vegas, NV
26 January 2025